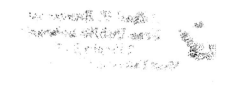

Feel the NOISE!

Chicago, Illinois

 Raintree

© 2006 Raintree
Published by Raintree,
A division of Reed Elsevier, Inc.
Chicago, Illinois

Customer Service   888–454–2279

Visit our website at www.raintreelibrary.com

Printed and bound in the United States by Lake Book Manufacturing, Inc

10 09 08 07 06
10 9 8 7 6 5 4 3 2 1

Library of Congress Cataloging-in-Publication Data
Claybourne, Anna.
  Feel the noise! : Electricity and circuits / Anna Claybourne.
      p. cm. -- (Fusion)
  Includes bibliographical references and index.
  ISBN 1-4109-1917-X (lib. bdg. : alk. paper) -- ISBN 1-4109-1948-X
(pbk. : alk. paper)
  1. Sound--Juvenile literature. 2. Vibration--Juvenile literature.
  I.
Title. II. Series: Fusion (Chicago, Ill.)
  QC225.5.C55 2005
  534--dc22
                                 2005012127

Acknowledgments
The author and publishers are grateful to the following for permission to reproduce copyright material: Alamy/ImageState (Robert Llewellyn) pp.20–21; Alamy/Redferns pp. 4–5; Alamy/Rubber Ball Productions pp. 16–17; Alamy/Waring Abbott p. 19 top; Powerstock/mjhunt.com pp. 26–27; Redferns (Phil Dent) pp. 10–11; Rex (Alex Maguire) pp. 8–9; Rex/Mephisto p. 19 bottom; Topfoto (Clive Barda)/PAL pp. 7, 11 inset; Topfoto/Photri p. 25; Topfoto (Talula Sheppard)/Arena PAL pp. 12–13.

Cover photograph of a performer singing with a band on stage, reproduced with permission of Lebrecht/Kaloyan Karageorgiev.

The publishers would like to thank Nancy Harris and Harold Pratt for their assistance in the preparation of this book.

Every effort has been made to contact copyright holders of any material reproduced in this book. Any omissions will be rectified in subsequent printings if notice is given to the publishers.

The paper used to print this book comes from sustainable resources.

Disclaimer
All the Internet addresses (URLs) given in this book were valid at the time of going to press. However, due to the dynamic nature of the Internet, some addresses may have changed, or sites may have changed or ceased to exist since publication. While the author and publishers regret any inconvenience this may cause readers, no responsibility for any such changes can be accepted by either the author or the publishers.

Contents

Some words are printed in bold, **like this**. You can find out what they mean on page 30. You can also look in the box at the bottom of the page where they first appear.

Meet the Band

A famous rock band is on tour. Millions of fans want to hear the band play. It needs to play in really big places like this sports stadium to fit everyone.

When the band members start playing, everyone around will hear them. It will be even more exciting for the fans in the stadium. The sound will be so loud that the fans will feel it shaking their whole bodies!

You know who your favorite bands are and why you love their music. But how does music—or any sound—actually work? What are sounds made of?

Turn the page
to feel the noise!

Setting Up

Before a show can start, the band members have to test their gear. The drummer starts by hitting a drum to check the sound. A drum is made of a tight "skin," or plastic, stretched across the end of a hollow box. When the stick hits the skin, the skin **vibrates**. It moves backward and forward many times a second. This makes a sound.

Sound is a type of **energy**. Energy is the ability to do some kind of work. This could be making things move. Sound is made when objects vibrate, or move backward and forward. All sounds are caused by vibrating objects.

Sound fact!

A bass drum skin vibrates around 250 times per second.

energy   something we use to do work
vibrate   to move backward and forward quickly

▼ When a stick hits a drum, it makes a noise. The sound happens because the stick makes the drum skin vibrate.

7

How does sound get to your ears?

Sound check

The band's **sound engineer** checks the instruments. He checks to hear if they are loud enough and if they sound good together. But how does the sound the band makes reach the sound engineer's ears?

The answer is that sound travels through air. When an object such as a drum **vibrates**, it pushes against the air around it. This starts **vibrations** in the air. These are called **sound waves**.

Sound waves spread out through the air and hit other objects. When they hit someone's ears, that person hears the sound.

The sound engineer checks▶ the sound of the instruments. He checks to hear if they are loud enough and if they sound good together.

sound engineer	sound expert who makes sure bands sound good
sound waves	sound vibrations moving through the air
vibrations	fast backward and forward movements

Feeling the noise

The band has people called roadies to set up all the gear. While the band is getting ready, the roadies carry things onto the stage. A roadie walks in front of one of the giant speakers. Suddenly, a "wall" of sound hits her.

Have you ever been somewhere really loud? If so, you might have felt sound hitting you, too. In fact, **sound waves** actually do hit you. They make the air move backward and forward. If the sound is very loud, the air moves so much that you can feel it.

To test this, hold a blown-up balloon against a radio or stereo speaker. Your fingers will feel the sound waves making the balloon **vibrate**.

percussion   instruments you play by hitting them

Sound fact!

Evelyn Glennie is deaf, but she is a famous **percussion** player. Evelyn always plays in bare feet. She feels the sound waves with her feet, face, and head.

Ready to Rock!

The stadium doors open. The fans rush in to get the best places near the front. The stadium goes dark. Then, the stage lights up and the band is on!

A huge cheer goes up as the band starts to play. The music is so loud that people living nearby can hear it, too. Some open their windows so that they can get a free concert.

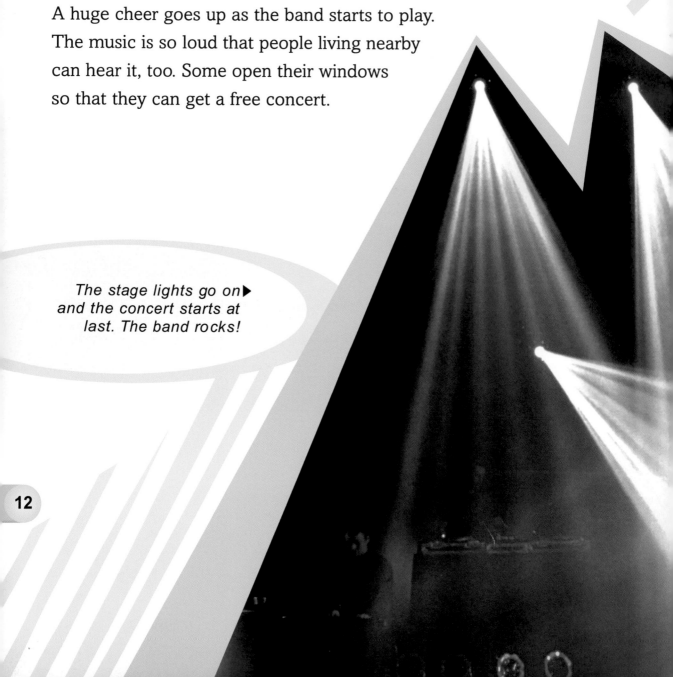

The stage lights go on▶ and the concert starts at last. The band rocks!

Sounds range from very quiet to very loud. You can make sound louder or quieter by using the controls on your radio, TV, or CD player.

So, how loud is really loud? Turn the page to find out!

How loud is that?

The loudness of a sound depends on how strong the sound **vibrations** are.

If you hit a drum gently, the skin only moves backward and forward and **vibrates** slightly. It sends slight vibrations into the air. If you hit it hard, the vibrations get bigger. The air vibrates more, too. The vibrations hit your ears harder, and so you hear a louder sound. Loudness is measured in units called **decibels** (dB).

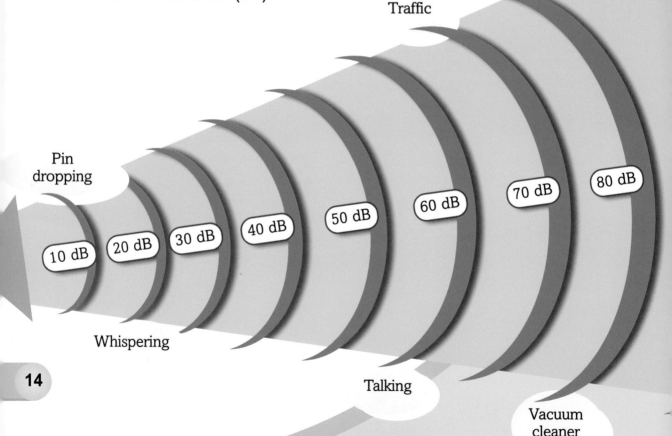

Lawnmower

Traffic

Pin dropping

Whispering

Talking

Vacuum cleaner

10 dB 20 dB 30 dB 40 dB 50 dB 60 dB 70 dB 80 dB

decibel (dB)	measurement of loudness
eardrum	thin skin in the ear that vibrates when sounds reach it
tinnitus	ringing or buzzing sound in the ears

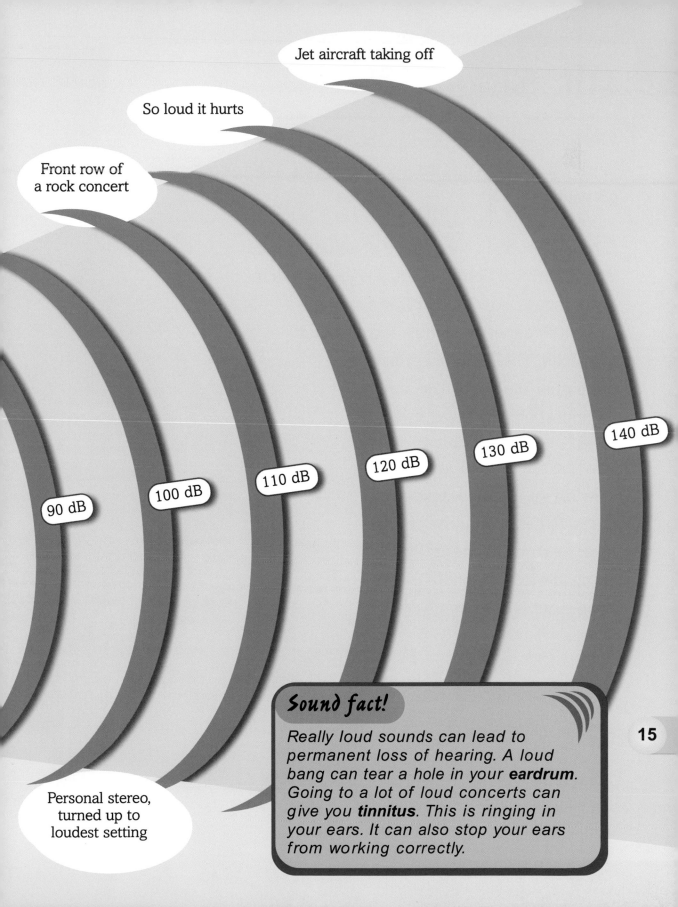

Hit Single

The band is halfway through the show. It plays its latest hit single. The fans know the song. They go wild!

A tune or song is a sequence of different notes. The difference between notes is called **pitch**. Pitch means how high or low a sound is. The pitch depends on how fast a sound **vibrates**. The faster the **vibration**, the higher the pitch. The slower the vibration, the lower the pitch.

You use the keys or strings of an instrument to make different notes. When a guitarist presses a guitar string, it makes the string shorter. The note that comes out is higher.

pitch   how high or low a sound is

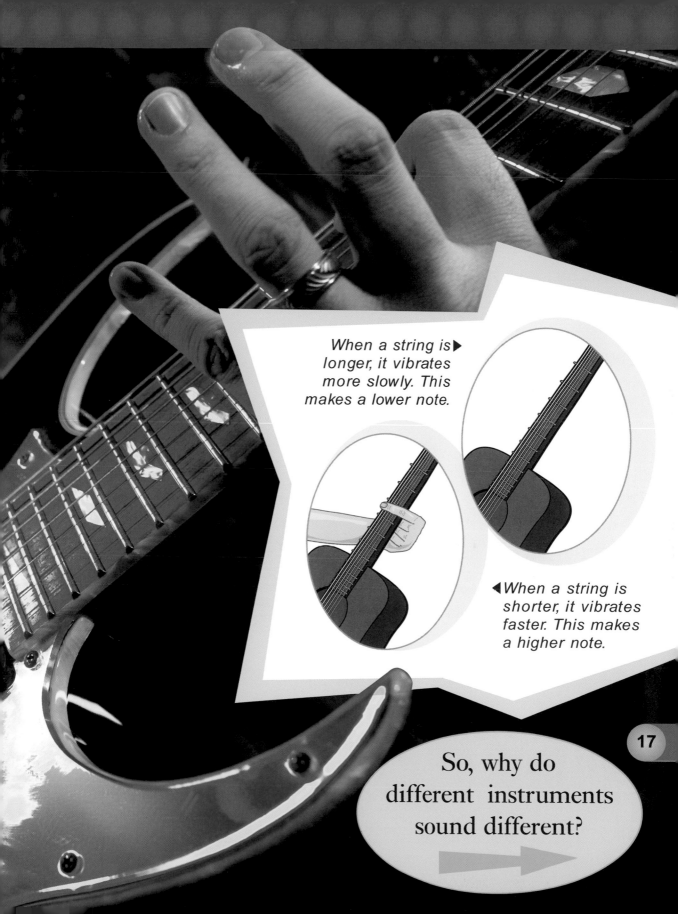

When a string is ▶ longer, it vibrates more slowly. This makes a lower note.

◀ When a string is shorter, it vibrates faster. This makes a higher note.

So, why do different instruments sound different?

Making music

Why do different parts of a band sound so good together?

Each instrument has its own special sound. Why? Because it makes its own patterns of **sound waves**. These patterns depend on the shape of the instrument and what it is made of. You can tell a trumpet from a guitar just by listening, even if both are playing the same note.

The sound waves of all the different instruments join together to make a song.

These graphs ▶ show the sound wave patterns each instrument makes. ▼

Electric Guitar

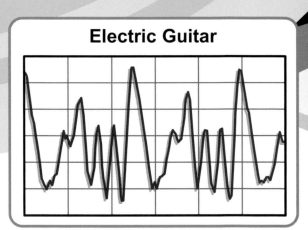

Singer

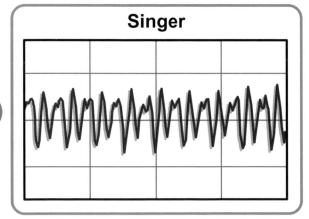

Trumpet

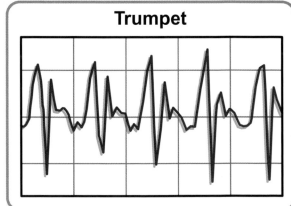

▲ A guitar works by making strings **vibrate** to make a sound.

You play a trumpet by making▼ your lips vibrate into it.

What about the singer?

The big chorus

The singer belts out the song at the top of her voice.
The fans are singing along, too.

The human voice is like an instrument. It changes **pitch**
to make different high and low notes. Inside your throat are
muscles called **vocal cords**. Your breath blowing past
them makes them **vibrate**.

For higher notes, your throat makes your vocal cords shorter.
This mean they vibrate faster. For lower notes, the vocal
cords get longer. This means they vibrate more slowly.

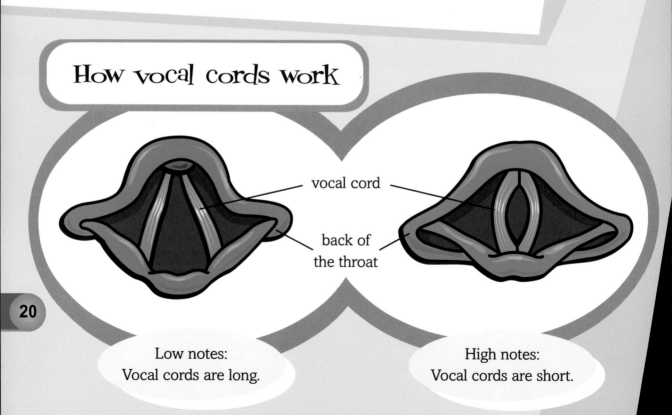

How vocal cords work

vocal cord

back of
the throat

Low notes:
Vocal cords are long.

High notes:
Vocal cords are short.

vocal cords   muscles in the back of the throat that make the voice work

The singer is often ▶ the most important person in a band. The singer sings the words and the main tune.

Sound fact!

Changing the pitch of your voice helps get your message across when you talk. Try saying "Yeah!" first in an excited way and then in a bored way. You use pitch to make them different.

21

How does hearing work?

Sounds Good

The band sounds better than ever! But how do you know what sounds good or bad? It is because of the way your ears and brain work together to help you hear.

The outer part of your ear catches sound **vibrations** in the air. The vibrations travel up a tube called the **ear canal.** Then, they hit your **eardrum**.

From there, the vibrations travel into a snail-shaped, bony area. This is called the **cochlea**. The cochlea turns the vibrations into sound signals. The sound signals travel to your brain. Then, your brain figures out what you are hearing.

Sound fact!

Many animals can hear better than humans. Scientists think dolphins may have the best hearing of all.

cochlea   snail-shaped part inside the ear
ear canal   hole leading into the ear

How you hear sounds

6

The brain figures out what you are hearing.

1

Sound waves

5

Sound signals travel into the brain along this nerve.

outer ear

ear drum

ear canal

4

The cochlea turns vibrations into sound signals.

cochlea

3

Vibrations travel up the ear canal and hit the eardrum.

2

The outer ear catches sound vibrations.

23

The Speed of Sound

The band is playing the last song. Everyone in the area can hear it. Did you know that the farther away the people are, the longer it takes for them to hear the sound?

If you were in the apartments just outside the stadium, you would hear the sound about a second after the band played it. If you were about half a mile (1 kilometer) away, you would hear the sound three seconds later. This is because **sound waves** take almost three seconds to travel this distance.

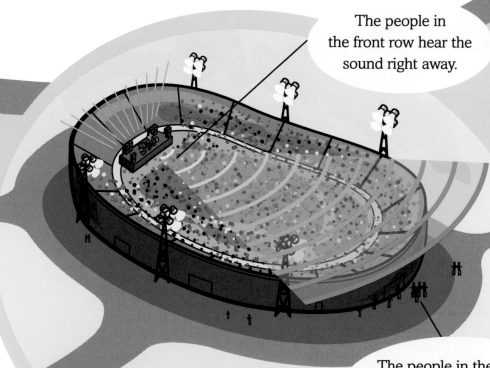

The people in the front row hear the sound right away.

The people in the street outside the stadium hear the sound a second later.

supersonic faster than the speed of sound

Sound fact!

If you watch a baseball game from a distance, you often hear the sound of a bat hitting a ball a moment after you see the actual hit. This is because it takes the sound a while to zoom through the air and reach your ears.

▲Some planes are **supersonic**. That means they can go faster than the speed of sound.

World of Sound

The show is over. The music stops. The fans leave. Yet there are still sounds everywhere—the sounds of everyday life.

Stop what you are doing right now and you will hear those sounds. You might hear traffic or the hum of a computer. Perhaps you can hear someone coughing or just your own breathing.

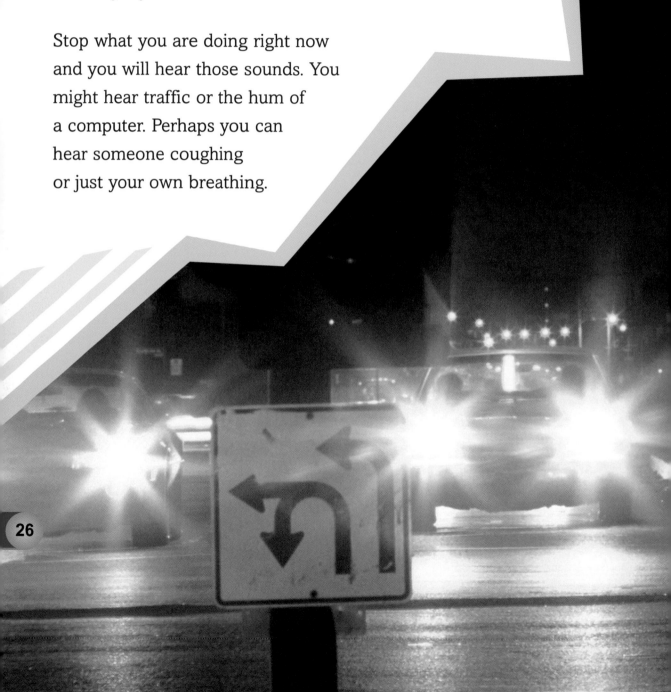

There is hardly a moment in your life when you can hear absolutely nothing. Many sounds, like a phone ringing or a fire alarm, can be really useful.

Just think how different life would be without sound . . .

As people leave the concert, there are many noises in the street outside. We are so used to these everyday noises that we ◀ hardly notice them.

Sound fact!

A composer named John Cage wrote a piece of music called 4'33". It is just 4 minutes, 33 seconds of silence! Cage wanted people to listen to the everyday sounds around them when the piece was performed.

Sound Facts

Remember that loudness is measured in **decibels** (dB).
For every 10 decibels you go up, sound gets ten
times as loud. So, a 50-decibel sound is 10
times louder than a 40-decibel sound.

The sound in the
front row of a rock
concert would measure
110 decibels

110 dB

100 dB

90 dB

80 dB

70 dB

60 dB

50 dB

40 dB

30 dB

20 dB

10 dB

Normal talking
measures 50 decibels.

Pitch facts

◀ **Pitch** *means how high or low a sound is. It is measured in Hertz (Hz).*

◀ *1 Hertz = 1* **vibration** *per second.*

◀ *A very low sound, such as a bass drum, has a pitch of about 250 Hertz.*

◀ *A high sound, such as a very high scream, has a pitch of about 3,000 Hertz.*

◀ *Humans can hear sounds between about 20 Hertz and 20,000 Hertz.*

There is no maximum on the decibel scale.

At 160 decibels, your **eardrums** would burst.

160 dB

150 dB

140 dB

130 dB

120 dB

Speed of sound facts

◀ *The speed of sound is measured in miles per hour (mph) or feet per second (ft/s).*

◀ *The Mach number is the speed of an aircraft compared to the speed of sound.*

   Mach 1 = the speed of sound
   Mach 2 = twice the speed of sound
   Mach 3 = three times the speed of sound

◀ *In 2004 the unmanned NASA jet plane X-43A flew at Mach 9.8, or nearly ten times the speed of sound. This is the fastest a plane has ever flown.*

29

Glossary

cochlea snail-shaped part inside the ear. The cochlea changes sound vibrations into nerve signals.

decibel (dB) measurement of loudness. A loud rock concert measures about 110 decibels.

ear canal hole leading into the ear. Sound travels along it to the eardrum.

eardrum thin skin in the ear that vibrates when sounds reach it. Very loud sounds can break your eardrum.

energy something we use to do work. Electricity, movement, and sound are all forms of energy.

percussion instruments you play by hitting them. Percussion instruments include drums, cowbells, and xylophones.

pitch how high or low a sound is. A "high-pitched" voice is a high, squeaky voice.

sound engineer sound expert who makes sure bands sound good. He or she checks the sound before a concert.

sound waves sound vibrations moving through the air. Unlike ocean waves, they are invisible.

supersonic faster than the speed of sound. Some planes are supersonic.

tinnitus ringing or buzzing sound in the ears. It can be caused by hearing too many loud noises.

vibrate to move backward and forward quickly. A guitar string vibrates when you pluck it.

vibrations fast backward and forward movements. All sounds are caused by vibrations.

vocal cords muscles in the back of the throat that make the voice work. You blow air past them to make a noise.

Want to Know More?

Books to read

• Cooper, Christopher. *Sound: From Whisper to Rock Band*. Chicago: Heinemann Library, 2004.

• Farndon, John. *Sound and Hearing*. Tarrytown, N.Y.: Marshall Cavendish, 2000.

• Parker, Steve. *Making Waves: Sound*. Chicago: Heinemann Library, 2005.

Websites

• http://www.sci.mus.mn.us/sound
Learn more about the science of sound at this cool site sponsored by the Science Museum of Minnesota.

•http://library.thinkquest.org/5116
Learn more about sound energy and what it looks like when it is recorded. Also find information about different instruments that play in an orchestra.

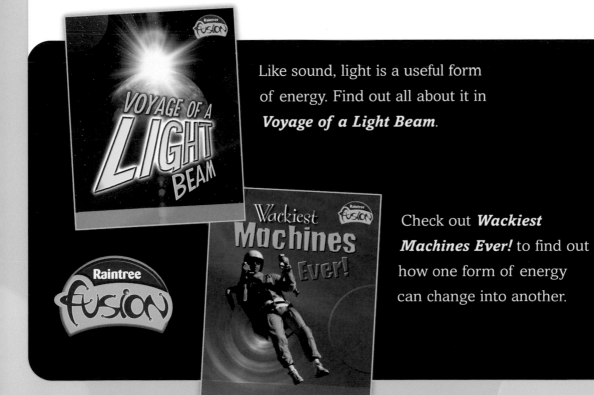

Like sound, light is a useful form of energy. Find out all about it in *Voyage of a Light Beam*.

Check out *Wackiest Machines Ever!* to find out how one form of energy can change into another.

Index